Problem Solving with Polyhedra Dice

Nancy Segal Janes

Cuisenaire Company of America, Inc.
White Plains, New York

Acknowledgments

Some of the games in this book have appeared in *Wonderful Ideas*, a newsletter featuring ideas for teaching, learning, and enjoying mathematics. Over the years, many teachers and students have contributed their wonderful ideas toward the development of these games and activities. While too numerous to thank individually, I would like to acknowledge, in particular, Rachel McAnallen and the Institute for Math Mania, Debby Shaw, Elizabeth Segal, Michael Kascak, Doris Hirschhorn, Clare Siska, Marilyn Sweeney, Paul Just, and Lynn Jurenka. And, a special thanks to George and Sophia.

Design and Cover: Arthur Celedonia
Production: Joshua Berger

ISBN 0-938587-74-9

2 3 4 5 6 – VG - 02 01 00 99 98

CONTENTS

Like many other teachers, I find that quality educational games have value and benefits in the classroom. Students stay focused on the activity, there is plenty of energetic, productive discussion, and learning occurs in a fun and motivating environment. With this in mind, I created *Problem Solving with Polyhedra Dice*, a collection of enjoyable and thought-provoking dice games and activities designed to link problem solving with math concepts taught in upper-elementary and middle-school classrooms.

The games use a variety of polyhedra dice, such as hexahedrons, octahedrons, and dodecahedrons. I like to use the proper mathematical terms for the specific polyhedrons instead of the word "dice" to stress the mathematical nature of the materials—and it's a nice way to help students learn the correct mathematical language. Students enjoy using the various polyhedrons and, often, the element of chance involved in polyhedra games allows those students who do not typically shine in math to be winners.

In addition to polyhedra dice, the following materials are needed for the activities in this book: operation dice (addition/subtraction and multiplication/division), fraction die, and small markers such as beans or counters.

Organization of the Book

Problem Solving with Polyhedra Dice has two parts. The first part contains 20 pupil activity pages, each backed by a playing board. Each pupil page has three sections:

> YOU NEED lists the materials needed for the activity.

> RULES provides information on who is the winner, how to play, and sample turns, moves, or rounds.

> WORKING TOGETHER often asks students to discuss their strategies for winning the games.

The second part of the book contains teacher notes for each activity. These notes include the math skills needed to play a game, suggestions on how to get students started, comments on the *Working Together* section, variations, and ideas for going further with a game.

Classroom Management

Using polyhedra dice—or any math manipulative—for the first time involves some management issues for teachers. I recommend playing the games beforehand so that you can begin to think about the strategies and thinking processes involved in a game. You can use class time to observe students and listen to their conversations as they work through strategies and challenging problems. New *Working Together* questions may occur to you while observing your students. Add these questions to the discussion, even if you don't know the answer; sometimes the best questions are those without answers.

Be sure to allow students plenty of time to explore the polyhedrons. Most students seem fascinated by the different polyhedrons and enjoy taking a closer look at them. Focusing their attention on the games later on will be easier if initially they are given time to explore the materials on their own.

In my classroom, students generally play the games at their desks or tables. To keep the polyhedrons from flying all over the room, I sometimes set a rule that if the polyhedrons fall onto the floor, the student loses a turn. If you want to keep the noise level down, have students roll the polyhedrons into the plastic supermarket meat trays.

Most games work best when students play as two-person team. Interacting with a partner encourages students to discuss and reflect on their strategies and ideas—a wonderful way for students to communicate mathematically.

NCTM Standards and Problem Solving

The 1989 National Council of Teachers of Mathematics *Curriculum and Evaluation Standards for School Mathematics* urged that four themes permeate all math curricula and instruction: problem solving, communication, reasoning, and connections. These themes are not separate concepts but are interrelated. They are incorporated into the games and activities in this book.

Mathematics learning needs to go beyond rote computation because the process of problem solving is more important than finding the exact solution. This problem solving should include routine and nonroutine problems so that students can test out their ideas by playing around with mathematics and numbers. One way to focus more on the process of problem solving is to provide calculators for students. This permits students to concentrate on the problem itself, not the computations. I recommend that as students explore the activities, you allow them to use calculators wherever appropriate.

Problem solving inevitably involves reasoning. A math curriculum that promotes reasoning is one that asks "Why" and "What if" questions. It also encourages students to explore and follow through on their ideas in situations in which there is more than one solution and more than one way to reach a solution. Students should have the opportunity to play each game several times so as to develop strategies. In some games they will be asked to compare their results with those of other teams.

Whereas seasoned mathematicians make connections between different topic areas all the time, young mathematicians tend to view the subject as as set of clearly defined and segregated topics. When students begin to realize that strategies, patterns, and methods used to solve one problem can be used to tackle a different problem, they begin to perceive mathematics as a connected whole. As students engage in these dice games and activities, they will find that strategies that work for one game can be used to help them win another one. For example, the same strategies can be used for winning **Turning Products**, a multiplication-based game, and **Gridlock**, a coordinate graphing game. Students who study probability in the **Dice Races** will discover that knowledge helpful for developing a strategy for games like **Pig Out** and **Position Power**.

It is also important for students to be able to communicate their discoveries, strategies, and thoughts about mathematics. To achieve this goal, provide students with opportunities to write, read, and discuss so that they can improve their abilities to express ideas clearly. Benefits of these games are derived from the discussions and analyses that occur during and after the game. Having students play the game in two-person teams encourages these discussions.

To further enhance problem solving, the *Working Together* section includes a rich variety of questions, many requiring explanations for answers. Many of these thought–provoking questions emphasize inquiring into the strategies and sharing the thinking processes students employed in playing the games. The questions tackle a variety of concepts, strengthening students' perceptions of math as a group of interrelated ideas. For example, in the **Dice Races**, students explore the probability of rolling sums and products with different pairs of polyhedrons. In **Reach for Fifty, Make It Close**, and **Fraction Attraction**, students explore estimation strategies.

Discussion in the Classroom

Classroom discussion is a very effective way to promote mathematical communication. When students explain their strategies for solving a problem, they must clarify their ideas in order to express them in words.

Leading a successful classroom discussion can be challenging. Here are a few tips I've picked up over the years:

- Allow students to share their ideas in a small group before having a whole-class discussion. This enables more students to be involved and provides a chance for them to clarify their thoughts before speaking to the entire group. Most of the activities in this book are set up to be played in small groups; those groups of students can collaborate to discuss the *Working Together* questions.

- Some students feel more comfortable speaking to a small group than to the whole class. One option is to have a group spokesperson who will summarize the group's responses before opening up a question for class discussion.

- After posing a question, give students 10 to 15 minutes to jot down a few notes in response. There always seem to be students who raise their hands before I even finish asking a question, whereas other students need a little more time to formulate their thoughts. If everyone is given a period of time to think about the question, more students will partake in the whole-class discussion.

- Always give the speaker enough time to finish responding and do not let other students (or yourself) interrupt the speaker.

- Keep in mind that when you ask students a question, it is because you are interested in their responses. After hearing an answer, many teachers repeat a student's exact words. Doing so takes away the importance from the student's thoughts and merely stresses the anwer. If you want the answer repeated because you are not sure if everyone heard or understood, either ask the student to repeat it so that everyone can hear, or ask a classmate to explain it in his or her own words.

- Writing students' names and responses on a large piece of paper can preserve the discussion long after it has been completed. This makes a nice bulletin board display, gives credit to small groups' or individuals' contributions, and enables you to remind students of important ideas brought up in previous classes.

Wonderful Ideas

A wonderful idea is an idea that comes from the student and is developed through active involvement with his or her learning. Eleanor Duckworth, author of *"The Having of Wonderful Ideas" and other Essays on Teaching and Learning* writes, "The more we help children to have their wonderful ideas and to feel good about themselves for having them, the more likely it is that they will some day happen upon wonderful ideas that no one else has happened upon before." Wonderful ideas come in many forms: a winning strategy for **Number Ladder**, a clever solution in **All the Way to Ten**, or a discovery about prime numbers in **Total Factor**.

There are some wonderful ideas out there in your classroom just waiting to be uncovered. Let the games begin!

Nancy Segal Janes

References

Duckworth, Eleanor. *"The Having of Wonderful Ideas" and other Essays on Teaching and Learning.* New York, N.Y.: Teachers College Press, 1987.

National Council of Teachers of Mathematics. *Curriculum and Evaluation Standards for School Mathematics.* Reston, Va.: NCTM, 1989.

Most of the games and activities in this book have an element of chance, or probability; which number will come up next on the roll of a polyhedron is always unknown. Probability is a way to measure the likelihood of an event occurring. This likelihood is used to predict and to make conjectures about future events.

Probability is the number of ways in which an event can occur divided by the total number of possible outcomes. This number is always between 0 and 1, or equal to 0 or to 1.

There are two types of probabilities: experimental and theoretical. Experimental probability involves performing an experiment, collecting data, and analyzing the outcome to predict what is likely to happen in similar events. Theoretical probability is based on knowing all the possible outcomes and determining probabilities based on this knowledge. Often in real-life situations, we do not know the theoretical probabilities, so we base our predictions on our previous experiences. For example, the weather forecast may predict a 60 percent chance of rain. This probability is based on the weatherperson's collected information and previous experience. There is no way to know the actual probability of rain.

When rolling polyhedrons, all possible outcomes are known, so both experimental and theoretical probabilities can be integrated into the activities. In some of the activities, you and your students will see that the actual rolls of the polyhedrons come very close to the theoretical probabilities.

Probability with One Hexahedron

Assuming that the hexahedron is fair and balanced, each side of the cube has an equal chance of being rolled. There are six possible outcomes: 1, 2, 3, 4, 5, and 6. The probabilities of certain events can be figured in this manner.

The probability of rolling a 3:

There is one way to roll a 3, and there are six possible outcomes. The probability of rolling a 3 equals 1/6.

The probability of rolling an odd number:

There are 3 ways to roll an odd number (1, 3, 5) and there are 6 possible outcomes. The probability of rolling an odd number equals 3/6, or 1/2.

The probability of rolling an 8:

There are 0 ways to roll an 8, and there are 6 possible outcomes. The probability of rolling an 8 equals 0/6. An event with a probability of 0 is an impossible occurrence.

The probability of rolling a number less than 8:

There are 6 ways to roll a number less than 8 (1, 2, 3, 4, 5, 6) and there are 6 possible outcomes. The probability of rolling a number less than 8 equals 6/6, or 1. An event with a probability of 1 is a definite occurrence, a sure thing.

Each of the 6 possible outcomes of rolling a hexahedron (1, 2, 3, 4, 5, 6) can be referred to as a simple event. In a probability experiment, the sum of all the simple events must equal 1. For example, the probability of rolling 2 is 1/6, whereas the probability of *not* rolling 2 is 5/6. These two events include all possible outcomes, and their sum (1/6 + 5/6) equals 1.

Probability with Two Hexahedrons

One way to examine the theoretical probabilities of rolling different sums with two hexahedrons is to create an addition table.

+	1	2	3	4	5	6
1	2	3	4	5	6	7
2	3	4	5	6	7	8
3	4	5	6	7	8	9
4	5	6	7	8	9	10
5	6	7	8	9	10	11
6	7	8	9	10	11	12

The table shows the 36 different possible rolls. Often when students try to list all the rolls, they may not include a roll of 3 and 6 and a roll of 6 and 3 as different rolls. If students were to use two different-colored hexahedrons, they would see that rolling 6 on the red cube and 3 on the white cube is different from rolling 6 on the white cube and 3 on the red cube.

From the table, you can count the number of times each sum occurs. To find the probability of rolling each sum, divide the number of occurrences by the total number of possible rolls, 36. The most likely sum to occur is 7; it has a probability of 6/36 or 1/6. This means that for every 6 rolls, a sum of 7 can be expected to occur once. The table below shows the probabilities of each sum expressed in the fraction's simplest form.

Sum	2	3	4	5	6	7	8	9	10	11	12
Number of Occurences	1	2	3	4	5	6	5	4	3	2	1
Probability	1/36	1/18	1/12	1/9	5/36	1/6	5/36	1/9	1/12	1/18	1/36

According to probability theory, the more times the hexahedrons are rolled, the more likely it is that the experimental rolls will come to approximate the expected theoretical rolls. For example, in 36 rolls, one would expect a sum of 11 to occur two times; in 72 rolls, a sum of 11 four times; in 144 rolls, eight times. These numbers may not actually occur, but it is *likely* that our rolls will be close to these numbers.

Some Related Experiments

Roll Every Number

This experiment asks the question "How many rolls does it take to roll at least one of each of the six numbers?" Have students work in pairs and roll one hexahedron, recording the number rolled. They continue rolling until each number has been rolled once and then count the number of rolls that were needed. Students should repeat the experiment at least five times and then find the average number of rolls for each round.

Ask students to compare their average with other students' averages and then find the class average. Tell students that using a computer simulation, after 10,000 trials, it took on the average 14.7 rolls before all six numbers occurred. Have students compare their results with the computer results.

Pick a Number

In this experiment, students test the probability that any number from 1 to 6 will be rolled 1 out of 6 times. Students pick a number from 1 to 6 and roll the hexahedron until the number appears. They keep track of the number of rolls. Students repeat the experiment at least five times and find the average number of rolls. Again, they should compare results with those of other students and then find the class average. According to probability theory, the results should be on average about 6 rolls.

This experiment can also be extended to test the occurrence of doubles when rolling two hexahedrons. People tend to think of doubles as a rare roll. According to probability theory, however, doubles are expected to occur 1 out of every 6 rolls, or 1/6 of the time. Students can test whether this theory holds up to experimental data.

Probability is a topic that is easily accessible to students of all ages. Explorations into probability for elementary and middle-grade students should begin with activities that allow students to conduct actual experiments and then collect and analyze the resulting data. Polyhedra dice provide a wonderful forum for doing this, as students can test and investigate many areas of probability theory in a fun and motivating environment.

Tetrahedron   Hexahedron   Octahedron   Decahedron   Dodecahedron   Icosahedron

Dice Race 1

2 players

YOU NEED

☐ 1 dodecahedron, numbered 1-12
☐ 3-4 race charts

RULES

The first number whose row is completely filled wins.

1. Predict which number will win the race. Discuss and record your predictions.
2. One person rolls the dodecahedron. The other person marks an X in that number's row on the race chart.
3. Continue rolling the dodecahedron until one row is filled with Xs.

Play the game at least three times.

SAMPLE ROLLS

First roll: 9
Second roll: 7
Third roll: 9

6						
7	X					
8						
9	X	X				
10						
11						

WORKING TOGETHER

◆ Discuss the following: *What are the chances that 1 will win the next race you play? Why? What are the chances that 10 will win? Why?*
◆ Be prepared to explain your reasoning in a class discussion.

1										
2										
3										
4										
5										
6										
7										
8										
9										
10										
11										
12										

Dice Race 2

2 players

YOU NEED

- ☐ 2 decahedrons, numbered 0-9
- ☐ 2-3 race charts
- ☐ 1 class race chart

RULES

The first number whose row is completely filled wins.

1. Predict which number will win the race. Discuss and record your predictions.
2. One person rolls the decahedrons and adds the numbers. The other person marks an X in that number's row on the race chart.
3. Continue rolling the decahedrons until one row is filled with Xs.
4. On the class race chart mark an X in that number's row.

Play the game at least twice or until the class race chart has a winner.

SAMPLE ROLLS

First roll: 8 and 7. Mark 15.
Second roll: 4 and 6. Mark 10.
Third roll: 5 and 5. Mark 10.

9					
10	X	X			
11					
12					
13					
14					
15	X				

WORKING TOGETHER

◆ Discuss the following: *Which sum(s) occurred most often? Why do you think this happened? Which sum(s) occurred the least? Why do you think this happened?*

◆ Together, write a convincing argument and include a chart, diagram, or picture to explain your answer to the following: *If you conducted another dice race, which numbers do you think would win? Why?*

0										
1										
2										
3										
4										
5										
6										
7										
8										
9										
10										
11										
12										
13										
14										
15										
16										
17										
18										

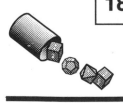

Dice Race 3

2 players

YOU NEED

☐ 2 octahedrons, numbered 1-8
☐ 3-4 race charts

RULES

The player with the most points wins.

1. Decide who will be the "even player" and who will be the "odd player." The even player scores points for all even products. The odd player scores points for all odd products.
2. One person rolls both octahedrons and multiplies the two numbers.
3. The other person marks an X in that number's row on the race chart.
4. Continue rolling the octahedrons until one row is filled with Xs.
5. Count the number of even products that were rolled; this is the even player's score. Count the number of odd products that were rolled; this is the odd player's score.

Play the game at least three times.

SAMPLE ROLLS

First roll: 9 and 1. Mark 9.
Second roll: 5 and 3. Mark 15.
Third roll: 3 and 3. Mark 9.

8							
9	X	X					
10							
12							
14							
15	X						
16							
18							

WORKING TOGETHER

◆ Discuss the following: *Is this a fair race? Why or why not? If not, how could you make it fair?*

◆ Be prepared to explain your answers in a class discussion.

1									
2									
3									
4									
5									
6									
7									
8									
9									
10									
12									
14									
15									
16									
18									
20									
21									
24									
25									
28									
30									
32									
35									
36									
40									
42									
48									
49									
56									
64									

Take Off!

2 players per team

2-4 teams

YOU NEED

☐ 2 octahedrons, numbered 1-8

☐ 1 playing board per team

☐ 15 small game pieces per team

RULES

The first team to take all the game pieces off the playing board wins.

1. Arrange 15 game pieces on any number on your playing board. You can put one piece on each number. Or you can put a few pieces on some numbers and no pieces on others.

2. Decide who goes first. Take turns.

3. Roll the dice and find the sum. Each team uses this sum.

4. If you have a game piece on the sum, remove it. You may remove only one piece per turn, even if you have more than one piece on that sum.

Play the game at least three times.

SAMPLE ROLLS

First roll: 2 and 6.

Both teams remove one game piece from 8.

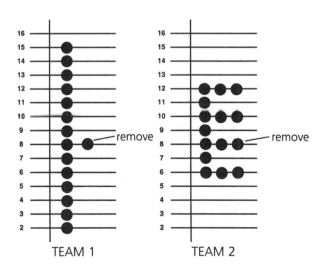

TEAM 1   TEAM 2

WORKING TOGETHER

◆ With your partner, draw a picture of the arrangement that you think works best for winning the game. Together, write a paragraph explaining why you think this arrangement is best.

◆ Compare your best arrangement with that of your opponents.

Take Off! BOARD

16 ——————————————————
15 ——————————————————
14 ——————————————————
13 ——————————————————
12 ——————————————————
11 ——————————————————
10 ——————————————————
9 ——————————————————
8 ——————————————————
7 ——————————————————
6 ——————————————————
5 ——————————————————
4 ——————————————————
3 ——————————————————
2 ——————————————————

Number Ladder

2 players per team
2-3 teams

YOU NEED

☐ 1 tetrahedron, numbered 1-4
☐ 2 hexahedrons, numbered 1-6
☐ 1 playing board per team

RULES

The first team to complete its Number Ladder in order from greatest to least wins.

1. Play *Game 1* using one tetrahedron and one hexahedron. Then play *Game 2* using two hexahedrons.
2. Decide who goes first. Take turns.
3. Roll the dice. Make a two-digit number with your roll.
4. Write the number in a box on your Number Ladder. From top to bottom, the numbers must be in order from greatest to least.
5. You may not be able to use every roll. For example, you may have already used the two-number combinations or there is no appropriate empty space.

Play the game at least twice.

SAMPLE ROLLS

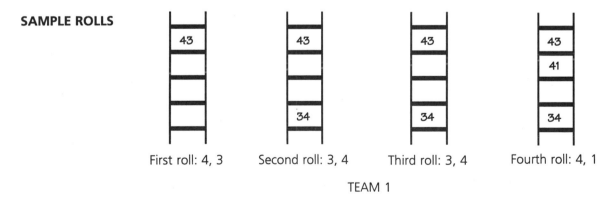

First roll: 4, 3   Second roll: 3, 4   Third roll: 3, 4   Fourth roll: 4, 1

TEAM 1

WORKING TOGETHER

◆ Make a list of all the two-digit numbers you can roll with each pair of dice.
◆ Discuss your strategies for winning. Make a list of all the strategies you come up with.

Number Ladder BOARD

Game 1   Game 2

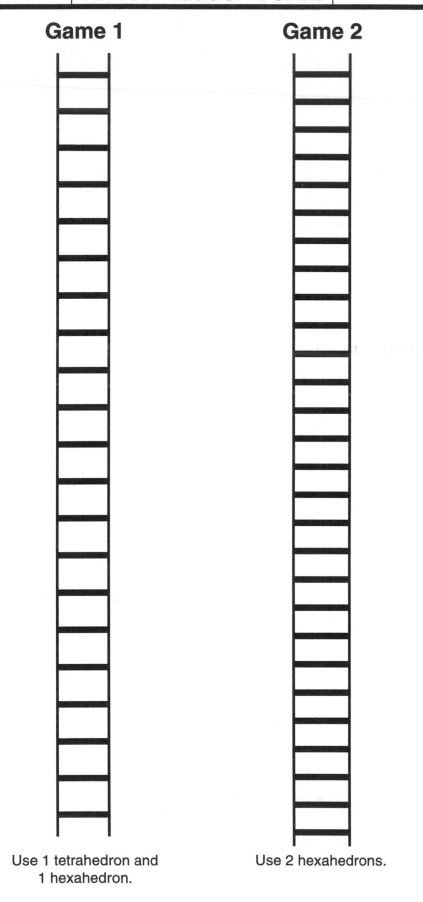

Use 1 tetrahedron and   Use 2 hexahedrons.
1 hexahedron.

Number Roll

2 players per team
2-3 teams

YOU NEED

☐ 1 decahedron, numbered 0-9
☐ 1 playing board per team

RULES

The team closest to the goal for that game wins.

1. Choose a game.
2. Roll the decahedron. Each team uses this number.
3. Write the number in one of the boxes.
4. Once you write a number in a box, you cannot move it.
5. You have one discard per game. Write this number in the discard box.
6. Take turns rolling the decahedron until all the boxes are filled.
7. Find the answer.
8. Repeat steps 2-7 for each of the other games.

Play each game at least twice.

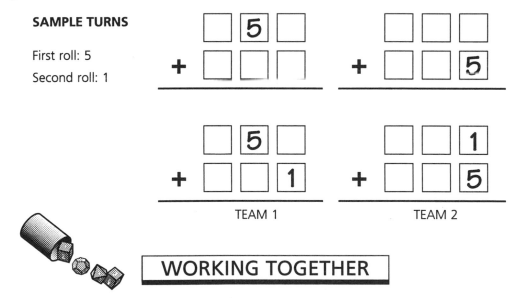

SAMPLE TURNS

First roll: 5
Second roll: 1

TEAM 1   TEAM 2

WORKING TOGETHER

◆ Discuss the following for each game: *If you rolled 9 on the first roll, where would you put it? Why? If you rolled 0 on the first roll, where would you put it? Why?*
◆ With your partner, rearrange the numbers for each game you played to determine the best answer. Compare your results with those of your opponents.

Game 1
Goal: Highest Sum

$$+ \;\; \square\;\square\;\square$$
$$\square\;\square\;\square$$

□
Discard

Game 2
Goal: Lowest Difference

$$- \;\; \square\;\square\;\square$$
$$\square\;\square\;\square$$

□
Discard

Game 3
Goal: Highest Product

$$\times \;\; \square\;\square$$
$$\square\;\square$$

□
Discard

Game 4
Goal: Lowest Quotient

$$\square \;\Big)\; \square\;\square\;\square$$

□
Discard

Game 5
Goal: Closest to 100

$$(\square + \square) \times (\square - \square) =$$

□
Discard

Place Your Values

2 players per team

2-3 teams

YOU NEED

☐ 1 decahedron, numbered 0-9
☐ 1 playing board per team

RULES

The team with the lowest total wins.

1. Roll the decahedron. Each team uses this number.
2. Write the number in one of the boxes on your playing board.
3. Once you write a number in a box, you cannot move it.
4. Take turns rolling the decahedron until all boxes are filled.
5. Find and record the difference between each target number and your number.
6. Find the total of the differences. This is your score for the game.

Play the game at least twice.

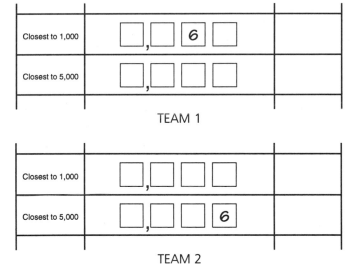

SAMPLE TURN

First roll: 6

WORKING TOGETHER

◆ Discuss the following: *Which place-value boxes did you try to fill first? Why?*
◆ Interview another team about the strategies it used.
◆ Write about the strategy you think is best.

Goal	Number	Difference
Closest to 100	☐ ☐	
Closest to 500	☐ ☐ ☐	
Closest to 1,000	☐,☐ ☐ ☐	
Closest to 5,000	☐,☐ ☐ ☐	
Closest to 10,000	☐ ☐,☐ ☐ ☐	
Closest to 100,000	☐ ☐ ☐,☐ ☐ ☐	
	Total	

Position Power

2 players per team

2-3 teams

YOU NEED

☐ 1 octahedron, numbered 1-8

☐ 1 playing board per team

RULES

The team closest to but not over 100 wins.

1. Decide who goes first. Take turns.
2. Roll the octahedron. On the first turn, write the number in one of the boxes for Round One. On your second turn, write the number in one of the boxes for Round Two.
3. Fill the empty boxes with zeros.
4. Find the sum of the two numbers.
5. Repeat steps 2-4 for three more rounds.

Play the game at least twice.

SAMPLE TURN

First roll: 2

Second roll: 8

Third roll: 4

Fourth roll: 5

```
  | 2 | 0 |   Round One
+ | 0 | 8 |   Round Two
----------
    2   8     Sum

+ | 4 | 0 |   Round Three
----------
    6   8     Sum

+ |   | 5 |   Round Four
```

TEAM 1

WORKING TOGETHER

◆ Discuss strategies you used. Then play again.

◆ If you could change the the positions of any of your numbers, what is the best score you could get? Compare your results.

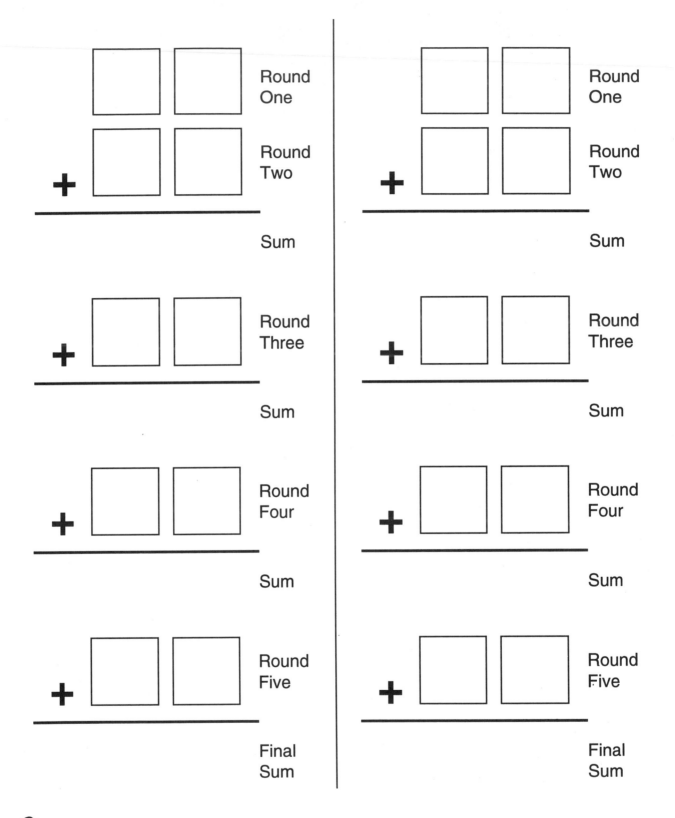

Round One

Round Two

$+$

Sum

$+$ Round Three

Sum

$+$ Round Four

Sum

$+$ Round Five

Final Sum

Round One

Round Two

$+$

Sum

$+$ Round Three

Sum

$+$ Round Four

Sum

$+$ Round Five

Final Sum

Place Your Decimals

2 players per team
2-3 teams

YOU NEED

- ☐ 1 decahedron, numbered 0-9
- ☐ 1 playing board per team

RULES

The team with the highest total score wins.

1. Roll the decahedron. Each team uses this number.
2. Write the number in one of the boxes on your playing board.
3. Once you write a number in a box, you cannot move it.
4. You have one discard per game. Write this number in the discard box.
5. Take turns rolling the decahedron until all the boxes are filled.
6. Score a point for each true math sentence. The sum of the points is your score for the game.

Play the game at least twice.

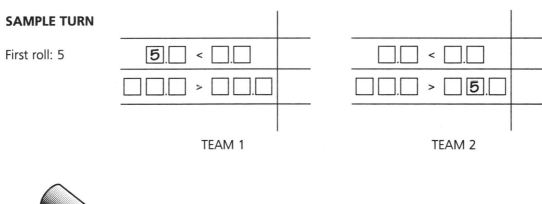

SAMPLE TURN

First roll: 5

TEAM 1   TEAM 2

WORKING TOGETHER

- ◆ Discuss the following: *Which place-value boxes did you try to fill in first? Why?*
- ◆ Interview another team about the strategies they used.
- ◆ Write about the strategy you think is best.

Decimal Inequalities	Points
□.□ < □.□	
□□.□ > □□.□	
.□□□ ≤ .□□□	
□.□□ + □.□ < □	
□.□□ − □.□□ > □	
Total	

□

Discard

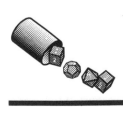

Wipe Out

2 players

YOU NEED

- ☐ 1 octahedron, numbered 1-8
- ☐ 1 hexahedron, numbered 1-6
- ☐ 1 playing board per player

RULES

The player with the highest score wins.

1. Decide who goes first.
2. Roll the octahedron and the hexahedron. Find the sum.
3. Cross off the sum or any other numbers on your playing board that add up to that sum.
4. Take turns rolling the dice until you cannot cross off any numbers. Your opponent continues until he/she cannot use a roll.
5. Find the sum of your crossed-off numbers. This is your game score.

Play the game at least three times.

SAMPLE TURN

First roll: 2, 4
Choose one.

Choice 1: 1 2 3 4 5 ⨯6 7 8 9

Choice 2: ⨯1 2 3 4 ⨯5 6 7 8 9

Choice 3: 1 ⨯2 3 ⨯4 5 6 7 8 9

Choice 4: ⨯1 ⨯2 ⨯3 4 5 6 7 8 9

WORKING TOGETHER

- ◆ Determine the lowest score possible.
- ◆ Create two different sequences of rolls that would enable you to cross off all 14 numbers.
- ◆ Find the fewest number of turns needed to wipe out all 14 numbers.
- ◆ Discuss the strategies you used to play the game. Write about the strategy you think is best.

Wipe Out BOARD

| 1 | 2 | 3 | 4 | 5 | 6 | 7 | 8 | 9 | 10 | 11 | 12 | 13 | 14 |

| 1 | 2 | 3 | 4 | 5 | 6 | 7 | 8 | 9 | 10 | 11 | 12 | 13 | 14 |

| 1 | 2 | 3 | 4 | 5 | 6 | 7 | 8 | 9 | 10 | 11 | 12 | 13 | 14 |

| 1 | 2 | 3 | 4 | 5 | 6 | 7 | 8 | 9 | 10 | 11 | 12 | 13 | 14 |

Pig Out

2 players

YOU NEED

☐ 2 octahedrons, numbered 1-8
☐ 1 score sheet per player

RULES

The first player to reach 100 wins.

1. Decide who goes first. Take turns.
2. Roll both octahedrons. Write the numbers and their sum on your score sheet.
3. If you choose, you can stop and end the round. Or, you can roll again and add the new sum to your previous score. But don't *pig out.* If you roll 1 on one octahedron, you lose all your points for that round. If you roll 1 on both octahedrons, your total score goes back to 0, even if your score is 99.
4. Any time you roll a 1, your turn is over.
5. Circle your score at the end of each round.
6. Take turns repeating steps 2-5.

Play the game at least three times.

SAMPLE ROUNDS

Points per roll	Roll	Score
7	3,4	7
10	4,6	17
4	2,2	(21) —— end of round 1
12	6,6	(33) —— end of round 2
7	3,4	40
0	1,5	(33) —— end of round 3
0	1,1	(0) —— end of round 4

WORKING TOGETHER

◆ Discuss the strategies you used. Write about the strategy you think is best.
◆ Make up a new rule for rolling 1 on both octahedrons. Play with your new rule. Discuss how the new rule changed your strategy.

Pig Out BOARD

Roll	Score	Roll	Score

Cross Out

2 players per team
2-3 teams

YOU NEED

☐ 2 octahedrons, numbered 1-8
☐ 1 playing board per team

RULES

The first team to reach 0 wins.

1. Each team chooses a starting number between 40 and 60.
2. Decide who goes first.
3. Roll both octahedrons.
4. Cross out one of the numbers *or* both of the numbers *or* the sum of the two numbers rolled.
5. Subtract the number(s) you crossed out from your starting number.
6. You may not be able to use every roll because you have already crossed out the number.
7. Take turns. Repeat steps 3-4.

Play the game at least three times.

SAMPLE TURN

Starting number: 45
First roll: 6, 2

Choice 1

1	2	3	4	5	6	7	8	9	10
1	2	3	4	5	6	7	8	9	10

Next turn: Subtract from 43

Choice 2

1	2	3	4	5	6	7	8	9	10
1	2	3	4	5	6	7	8	9	10

Next turn: Subtract from 39

Choice 3

1	2	3	4	5	6	7	8	9	10
1	2	3	4	5	6	7	8	9	10

Next turn: Subtract from 37

Choice 4

1	2	3	4	5	6	7	8	9	10
1	2	3	4	5	6	7	8	9	10

Next turn: Subtract from 37

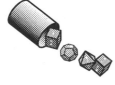

WORKING TOGETHER

◆ Discuss the following: *Which numbers make good starting numbers? Why? Which numbers did you cross out first? Why? Did you try to save some numbers for last? Which ones? Why?*

◆ Interview another team about the strategies it used.

◆ Write about the strategy you think is best.

1	2	3	4	5	6	7	8	9	10
1	2	3	4	5	6	7	8	9	10

Starting
Number

1	2	3	4	5	6	7	8	9	10
1	2	3	4	5	6	7	8	9	10

Starting
Number

Turning Products

2 players per team

2 teams

YOU NEED

- ☐ 2 dodecahedrons, numbered 1-12
- ☐ 2 different-colored markers, about 20 per team
- ☐ 1 playing board

RULES

The first team to get four markers in a row–horizontally, vertically, or diagonally–wins.

1. Decide who goes first. Take turns.
2. Roll both dodecahedrons, multiply the numbers, and place a marker on the product.
   NOTE: After the first roll, the dice are *turned*, not rolled.
3. Your opponents *turn* one of the dodecahedrons to a new number, multiply the two numbers, and place a marker on the new product.
4. Once a product is covered, it cannot be used again.
5. The five stars on the playing board are wild squares and do not get covered. They can be used by any team to help get four markers in a row.

Play the game at least three times.

SAMPLE TURNS

Roll: 5, 3
Cover 15.

1	★	2	3	4	5	6	7
8	9	10	11	12	14	●	★
16	18	20	21	22	24	25	27
28	30	32	33	35	36	40	42
44	45	48	★	49	50	54	55
56	60	63	64	66	70	72	77
80	81	84	88	90	96	99	★
100	★	108	110	120	121	132	144

TEAM 1

Turn: 9,3
Cover 27.

1	★	2	3	4	5	6	7
8	9	10	11	12	14	●	★
16	18	20	21	22	24	25	●
28	30	32	33	35	36	40	42
44	45	48	★	49	50	54	55
56	60	63	64	66	70	72	77
80	81	84	88	90	96	99	★
100	★	108	110	120	121	132	144

TEAM 2

WORKING TOGETHER

- ◆ Discuss winning strategies for this game.
- ◆ Write about the strategy you think is best.

1		2	3	4	5	6	7
8	9	10	11	12	14	15	⭐
16	18	20	21	22	24	25	27
28	30	32	33	35	36	40	42
44	45	48	⭐	49	50	54	55
56	60	63	64	66	70	72	77
80	81	84	88	90	96	99	⭐
100	⭐	108	110	120	121	132	144

Dodecahedron 1   Dodecahedron 2

All the Way to Ten

2 players per team
2-3 teams

YOU NEED

- ☐ 1 octahedron, numbered 1-8
- ☐ 1 decahedron, numbered 0-9
- ☐ 2 hexahedrons, numbered 1-6
- ☐ 1 tetrahedron, numbered 1-4
- ☐ 1 playing board per team

RULES

The team with the most points after ten rounds wins.

1. Decide who goes first. Take turns.
2. Roll all five polyhedrons. Record the numbers in the boxes for Round 1.
3. Use as many of the five numbers as you can and any mathematical operations or symbols to make an equation that equals the number of the round.
4. Write the equation on your playing board.
5. Score 1 point for each number you use and 1 bonus point if you use all five numbers.
6. Repeat steps 2-5 nine more times.
7. Find the sum of the points for each round. This is your game score.

Play the game at least twice.

SAMPLE TURNS

Round	Roll	Equation	Points
1	1 3 4 4 6	$14 - 6 - 4 - 3 = 1$	6

TEAM 1

Round	Roll	Equation	Points
1	1 3 6 8 9	$\dfrac{8 - 3 + 1}{6} = 1$	4

TEAM 2

WORKING TOGETHER

◆ Roll the five polyhedrons. Use the same five numbers to make equations equal to the numbers 1 through 10.

◆ Discuss your strategies for making equations. Play again.

Round	Roll	Equation	Points
1	☐ ☐ ☐ ☐ ☐		
2	☐ ☐ ☐ ☐ ☐		
3	☐ ☐ ☐ ☐ ☐		
4	☐ ☐ ☐ ☐ ☐		
5	☐ ☐ ☐ ☐ ☐		
6	☐ ☐ ☐ ☐ ☐		
7	☐ ☐ ☐ ☐ ☐		
8	☐ ☐ ☐ ☐ ☐		
9	☐ ☐ ☐ ☐ ☐		
10	☐ ☐ ☐ ☐ ☐		
		Total	

Reach for Fifty

2 players per team

2-3 teams

YOU NEED

☐ 2 hexahedrons, numbered 1-6

☐ 1 octahedron, numbered 1-8

☐ 1 addition/subtraction operation die

☐ 1 multiplication/division operation die

☐ 1 playing board per team

RULES

The team having the score closest to 50 after five rounds wins.

1. Roll the polyhedrons. Each team uses these numbers and symbols.

2. Create and solve an equation using two or three of the numbers and one operation. For division equations, round all quotients to the nearest whole number.

3. Write the equation on your playing board.

4. Take turns rolling the polyhedrons for four more rounds.

5. Find the sum of the answers to your equations. This is your game score.

Play the game at least twice.

SAMPLE ROUND

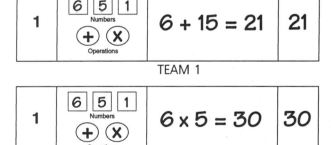

TEAM 1

TEAM 2

WORKING TOGETHER

◆ Discuss the strategies you used to play the game. Play again.

◆ Find five rolls that will give a game score of 50.

Problem Solving with Polyhedra Dice © 1994 Cuisenaire Company of America, Inc.

Reach for Fifty BOARD

Round	Roll	Equation	Answer
1	☐ ☐ ☐ Numbers ◯ ◯ Operations		
2	☐ ☐ ☐ Numbers ◯ ◯ Operations		
3	☐ ☐ ☐ Numbers ◯ ◯ Operations		
4	☐ ☐ ☐ Numbers ◯ ◯ Operations		
5	☐ ☐ ☐ Numbers ◯ ◯ Operations		

Total

Make It Close

2 players per team

2-3 teams

YOU NEED

- ☐ 1 icosahedron, numbered 1-20
- ☐ 5 polyhedrons (1 dodecahedron, 1 decahedron, 1 octahedron, 1 hexahedron, 1 tetrahedron)
- ☐ 1 addition/subtraction operation die
- ☐ 1 multiplication/division operation die
- ☐ 1 playing board per team

RULES

The team with the lowest score at the end of five rounds wins.

1. Roll the icosahedron to find your target number. Each team finds its own target number.
2. Roll the other seven dice. Each team uses these numbers and operations.
3. Make an equation that comes as close as possible to your target number. Use at least three of the numbers and at least one of the operations.
4. Find the difference between the solution to your equation and your target number. This is your score for the round.
5. Take turns rolling the dice. Repeat steps 1-4 four more times.
6. Find the total of the differences. This is your game score.

Play the game at least twice.

SAMPLE ROUND

Round	Target Number	Numbers and Operations Rolled	Equation	Difference
1	9	2,3,4,4,8 ÷, +	(32÷8)+4 = 8	1

TEAM 1

Round	Target Number	Numbers and Operations Rolled	Equation	Difference
1	18	2,3,4,4,8 ÷, +	48 ÷ 3 = 16	2

TEAM 2

WORKING TOGETHER

◆ Discuss the following: *Which numbers did you find most useful? Which operations? Why? Which numbers seemed to be rolled most often? Least often? Why do you think this happened?*

◆ Pick one of your rolls. Make equations that come as close as possible to each of the numbers 1-20. Compare your results.

Problem Solving with Polyhedra Dice © 1994 Cuisenaire Company of America, Inc.

Make It Close BOARD

Round	Target Number	Numbers and Operations Rolled	Equation	Difference
1				
2				
3				
4				
5				
			Total	

Total Factor

2 players per team
2-3 teams

YOU NEED

☐ 4 octahedrons, numbered 1-8
☐ 1 playing board per team

RULES

The first team to score a total of 400 points wins.

1. Decide who goes first. Take turns.
2. Roll all four octahedrons. Write the numbers on your playing board.
3. Find the sum of the four numbers.
4. List all the factors of the sum.
5. Add the factors. This is your score for the round.
6. Keep a running total of your scores.

Play the game at least twice.

SAMPLE TURNS

	Roll	Sum of Roll	Factors	Sum of Factors	Running Total
First turn	7 4 3 2	16	1, 2, 4, 8, 16	31	31
Second turn	5 3 2 1	11	1, 11	12	43

TEAM 1

WORKING TOGETHER

◆ Discuss the following: *Which sum gives you the lowest-scoring roll? Which sum gives you the highest-scoring roll? Can you score every number in between these two sums? How?*

◆ Make a list of all the numbers from 1 to 32. Find the numbers with only two factors. Describe them. Find the numbers with an odd number of factors. Describe them. Look for other patterns and describe them.

Total Factor BOARD

Roll	Sum of Roll	Factors	Sum of Factors	Running Total
☐ ☐ ☐ ☐				
☐ ☐ ☐ ☐				
☐ ☐ ☐ ☐				
☐ ☐ ☐ ☐				
☐ ☐ ☐ ☐				
☐ ☐ ☐ ☐				
☐ ☐ ☐ ☐				
☐ ☐ ☐ ☐				
☐ ☐ ☐ ☐				
☐ ☐ ☐ ☐				
☐ ☐ ☐ ☐				

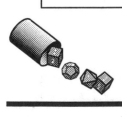

Fraction Attraction

2 players per team
2-3 teams

YOU NEED

☐ 3 decahedrons, numbered 0-9
☐ 1 playing board per team

RULES

The team with the sum closest to but not over 10 wins.

1. Decide who goes first. Take turns.
2. Roll the decahedrons. Choose two of the numbers and make a fraction.
3. Do this four more times.
4. When your playing board is filled, find the sum of your fractions. This is your game score.

Play the game at least twice.

SAMPLE ROUND

Round	Roll	Fraction
1	1 3 5	$\dfrac{5}{1}$

TEAM 1

Round	Roll	Fraction
1	0 1 8	$\dfrac{1}{8}$

TEAM 2

WORKING TOGETHER

◆ Discuss the following: *If you could change the fraction in one round, which round would it be? Why? What is your new fraction? How does your game score change?*

◆ Discuss your strategies for playing the game. Play the game again.

Problem Solving with Polyhedra Dice © 1994 Cuisenaire Company of America, Inc.

Round	Roll	Fraction
1	☐ ☐ ☐	
2	☐ ☐ ☐	
3	☐ ☐ ☐	
4	☐ ☐ ☐	
5	☐ ☐ ☐	
	Total	

Close Call

2 players per team
2-3 teams

YOU NEED

- ☐ 1 fraction die, numbered 1/12, 1/8, 1/6, 1/4, 1/3, 1/2
- ☐ 5 polyhedrons (1 dodecahedron, 1 decahedron, 1 octahedron, 1 hexahedron, 1 tetrahedron)
- ☐ 1 addition/subtraction operation die
- ☐ 1 playing board per team

RULES

The team with the lowest total score wins.

1. Roll the fraction die to find the target fraction. Each team uses the same target fraction.
2. Roll the remaining dice and record the roll. Each team repeats this step.
3. Using four of your numbers and the operation, make an equation with two fractions whose answer comes as close as possible to the target fraction.
4. Find the difference between your answer and the target fraction. This is your score for the round.
5. Repeat steps 1-4 four more times.
6. Total the differences. This is your game score.

Play the game at least twice.

SAMPLE ROUND

Round	Target Fraction	Roll	Equation	Difference
1	$\frac{1}{8}$	2 2 3 / 3 6 (−)	$\frac{2}{3} - \frac{3}{6} = \frac{1}{6}$	$\frac{1}{24}$

TEAM 1

Round	Target Fraction	Roll	Equation	Difference
1	$\frac{1}{8}$	1 1 3 / 6 6 (+)	$\frac{1}{6} + \frac{1}{6} = \frac{2}{6}$	$\frac{5}{24}$

TEAM 2

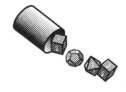

WORKING TOGETHER

◆ Discuss the strategies you used to play the game. Play again.
◆ Roll the number dice. Use these numbers and either operation to make six equations, each equal to one of the six fractions on the fraction die.

Close Call BOARD

Round	Target Fraction	Roll	Equation	Difference
1		☐ ☐ ☐ ☐ ☐ ◯		
2		☐ ☐ ☐ ☐ ☐ ◯		
3		☐ ☐ ☐ ☐ ☐ ◯		
4		☐ ☐ ☐ ☐ ☐ ◯		
5		☐ ☐ ☐ ☐ ☐ ◯		
			Total	

Gridlock

2 players

YOU NEED

☐ 1 blue decahedron, numbered 0-9
☐ 1 red decahedron, numbered 0-9
☐ game pieces in two different colors, about 30 each
☐ 1 playing board

RULES

The first player to get four markers in a row – horizontally, vertically, or diagonally – wins.

1. Decide who goes first.
2. Roll the decahedrons and place them on the playing board.
   The red decahedron shows the *x*-coordinate.
   The blue decahedron shows the *y*-coordinate.
3. Place a marker on the point shown by the two decahedrons.
4. Your opponent *turns* one of the decahedrons to create a new point and puts a marker on it.
   NOTE: After the first roll, the dice are *turned*, not rolled.
5. Take turns turning the decahedrons and placing your markers on the playing board.

Play the game at least three times.

SAMPLE MOVES

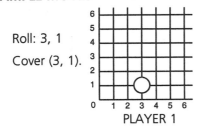

Roll: 3, 1

Cover (3, 1).

PLAYER 1

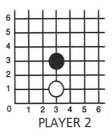

Turn blue
die to 3.

Cover (3, 3).

PLAYER 2

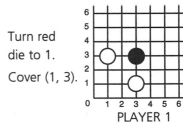

Turn red
die to 1.

Cover (1, 3).

PLAYER 1

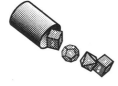

WORKING TOGETHER

◆ Discuss the strategies you used to play the game. Play again.
◆ Change the rule to three in a row wins. Play at least three times. Discuss how your strategy changed.

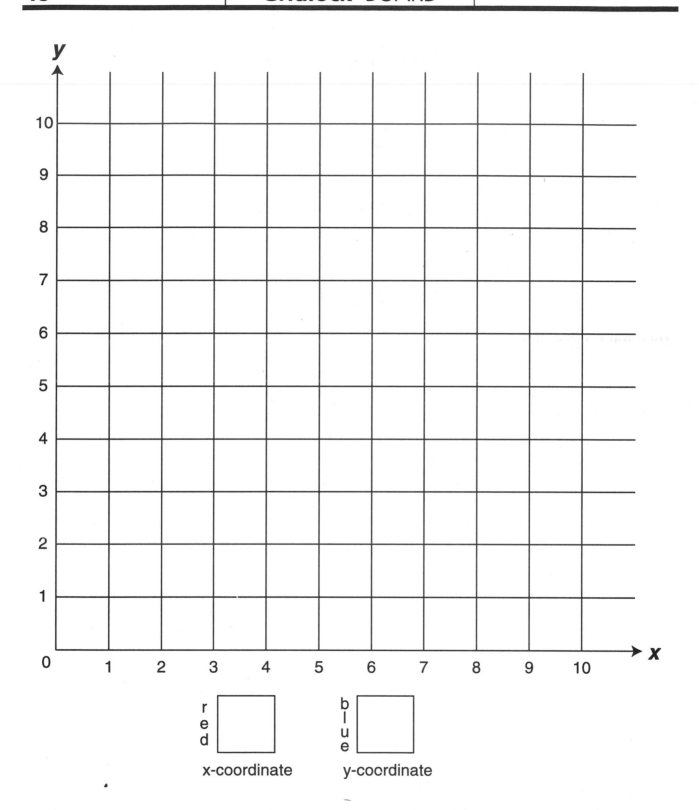

red x-coordinate

blue y-coordinate

Dice Race 1

2 players

MATH SKILLS

probability, equivalent fractions

THE GAME

Getting Started

Dice Race 1 is a simple experiment that introduces probability. Students run a race by rolling one dodecahedron. As students predict the winner, they begin to focus on the meaning of equally likely outcomes. Playing the game at least three times helps students to understand the value of using a large sample rather than a small one to make predictions.

Variations

Give students the following polyhedrons: tetrahedron, octahedron, decahedron, and hexahedron. Have students compare the polyhedrons and list ways in which they are alike and how they are different.

Have students play the game with a tetrahedron. Students will need a new race chart numbered 1-4. Ask students to predict the chances that 1 will win and to explain their reasoning.

Have students try the dice race with the remaining polyhedrons.

THE MATHEMATICS

Student Observations

Discussing students' predictions is an excellent way to discover what students know about probability. The first time students play **Dice Race 1**, they often select a favorite number or choose a number because "I always seem to roll that number."

When asked about their results, some students will say that 1 has just as good a chance as any other number but aren't sure why. Others will say that 1 has not won a race for them yet and is due to win. With experience, students' responses will begin to reflect their understanding that every number has the same chance of occurring, or a 1 out of 12 chance of winning.

Going Further

Discuss the probability of rolling each number on the dodecahedron; that each number has an equal chance of being rolled. For more information, see Notes on Probability.

Dice Race 2

2 players

MATH SKILLS

probability, addition, equivalent fractions

THE GAME

Getting Started

Although the math required to play **Dice Race 2** is simple, it sets the stage for an in-depth study of probability. In this race, students find the sum of two decahedrons and mark the sum on a race chart. The first sum to fill a row with Xs is the winner and is then marked on the class chart. The class chart should be a larger version or an overhead transparency of the race chart. On this chart, students will record the "winner" of each race.

Before distributing the playing boards, dis-cuss the sums that are possible with two dec-ahedrons. Ask for the lowest and the highest sums that can be rolled, and if all the sums between those numbers can be rolled.

Variations

Have students play the game with two tetra-hedrons. Ask students to predict the chances that 1 will win and to explain their reasoning.

Have them play the game with other polyhe-drons–hexahedrons, octahedrons, dodecahe-drons, and icosahedrons.

THE MATHEMATICS

Student Observations

Students' predictions provide insight into their understanding of probability. For students with little experience with probability, the first prediction is usually a favorite or lucky num-ber. Later predictions may be based on what they noticed while playing the game. They might say "We seemed to be rolling lots of 9s, so I guessed that." Or "Ten won the first two races, so I think it will win again." Students will likely report that 8, 9, and 10 seem to win the majority of races, with 9 usually winning most. After a few races, students generally have a sense that more combinations equal these numbers than equal the numbers at the end of the range, such as 0 and 18.

Going Further

To help them understand what they have observed, challenge students to determine how many different rolls are possible when rolling two decahedrons. This is different from asking how many sums are possible. For example, 1 and 6 is a different roll from 3 and 4 even though both have a sum of 7. Have stu-dents share the different ways they organize and solve this problem.

Help students to calculate the probability of each sum occurring. For information on cal-culating probabilities, see Notes on Probability.

Dice Race 3

2 players

MATH SKILLS

probability, addition, multiplication, factors, odd and even numbers, equivalent fractions

THE GAME

Getting Started

In **Dice Race 3,** two players roll two octahedrons and score points for odd or even products. Like the other dice races, the math required to play the game is simple. The discussion and analysis of the fairness of the race allow for a look at an important topic of probability–equally likely outcomes.

Before distributing the playing board, have students find the lowest and highest products possible using two octahedrons. Discuss whether it is possible to roll all the numbers between the two products. (No. For example, 11, 13, 17 are not possible.)

Variations

Have students find the sum instead of the product. Discuss with students if this is a fair race. If not, have them find ways to make it fair.

Challenge students to work in pairs to create their own fair and unfair polyhedron races, which can then be played and analyzed by their classmates.

THE MATHEMATICS

Student Observations

Students will most likely report that the game is unfair because the even player won more races than the odd player or because there are more even products than odd products on the race chart.

Students will have many suggestions for making the race fair. For example, one player scores points for products less than or equal to 18 and the other player scores points for products greater than 18. Or, the odd player scores 3 points for every odd product and the even player scores 1 point for an even product.

Going Further

Have students fill in an 8 x 8 multiplication chart and count the number of odd and even-products. Students should observe that they can expect to roll an odd product 16 out of 64

times, or 1 out of 4. They can expect to roll an even product 48 out of 64 times, or 3 out of 4. So the chance of rolling an even product is 3 to 1, giving an unfair advantage to the even player.

Challenge students to further explain their results by using an odd/even multiplication matrix like the one shown.

x	O	E
O	O	E
E	E	E

This matrix shows that there is a 1 in 4 chance of rolling an odd product and a 3 in 4 chance of rolling an even product. For more information on calculating probabilities, see Notes on Probability.

Take Off!

2 players per team
2-3 teams

MATH SKILLS

probability, addition

THE GAME

Getting Started

Take Off! is a strategy game based on the probability of rolling different sums with two octahedrons. Playing **Take Off!** does not require complex math.

Students distribute 15 game pieces on a playing board that resembles a number line from 2 to 16. The goal is to be the first team to remove all the pieces from the playing board.

Before students begin to play, discuss the sums that are possible when two octahedrons are rolled.

Variations

Have students play the game using other polyhedra dice. Students will need a new playing board for each different set of dice. Discuss with students how they changed their strategy for the dice they chose.

THE MATHEMATICS

Student Observations

Students will come up with different arrangements. Many will say that the game pieces should be focused around 8, 9, and 10 with a few pieces on the numbers at the ends of the board since the numbers 8, 9, and 10 are more likely to be rolled. Others might say that all the pieces should be evenly distributed since all the numbers are possible.

Going Further

Explore the probability behind rolling two octahedrons. Ask students to list all the sums that can be rolled and then ask them to list all the possible rolls for each of the sums. For example, 4 can be rolled in 3 ways, 1 and 4, 2 and 2, and 4 and 1. After discussing the list, pose this question: *If 8, 9, and 10 are most likely to be rolled, do you think it is a good strategy to put all your game pieces on those three numbers? Why?* For more information, see Notes on Probability.

Number Ladder

2 players per team
2-3 teams

MATH SKILLS

place value, estimation

THE GAME

Getting Started

In **Number Ladder**, students roll two polyhedrons to create a list of two-digit numbers in order from greatest to least. The goal is to be the first team to completely fill in the ladder. For students to place the numbers in order, they must estimate the best location for each number.

In *Game 1*, students use one tetrahedron and one hexahedron. In *Game 2*, students use two hexahedrons. The rules are the same for each game.

Variations

Have students play the game using three tetrahedrons. They will be able to form numbers from 111 and 444 containing the digits 1, 2, 3, and 4. Students will need a longer ladder for this game.

Have students make up their own playing boards. They can decide which polyhedrons to use and the shape of the boards.

In addition to the two polyhedrons, students can use a hexahedron labeled with three positive and three negative signs to introduce integers into the game. Again, they will need a longer ladder for this game.

THE MATHEMATICS

Student Observations

For *Game 1*, the possible rolls are:

11	21	31	41	51	61
12	22	32	42	52	62
13	23	33	43	53	63
14	24	34	44	54	64
15	25	35	45		
16	26	36	46		

For *Game 2*, the possible rolls are:

11	21	31	41	51	61
12	22	32	42	52	62
13	23	33	43	53	63
14	24	34	44	54	64
15	25	35	45	55	65
16	26	36	46	56	66

A strategy many students use is to fill the ends of the ladder *first* with high and low numbers. If the roll will not make a high or low number, they make their best estimate as where to locate the number.

Going Further

Discuss the probability of each roll. For example, have students compare the chance of rolling double 1s to the chance of rolling 3 and 2 or 5 and 4. Also, have students contrast the choices of a particular roll in *Game 1* with its chance of occurring in *Game 2*.

Have students play other games such as **Position Power** or **Place Your Values** for more practice in place value.

Number Roll

2 players per team
2-3 teams

MATH SKILLS

place value, addition, subtraction, multiplication, division

THE GAME

Getting Started

Number Roll provides students with practice in both arithmetic and problem solving at the same time. For example, in *Game 1*, students are given an addition problem. Using a decahedron, they fill in the boxes with the numbers rolled. The goal for this game is to find the problem with the highest sum. The goal changes for each game on the playing board. As students write the numbers, they must consider the place value of the number.

Variations

Number Roll can be used to practice many

different mathematical operations and concepts. For example, to have students practice specific multiplication facts, create examples with one number already inserted such as:

To have students practice fraction concepts, create inequalities such as these:

$$\frac{\square}{\square} < \frac{\square}{\square} \qquad \frac{\square}{\square} + \frac{\square}{\square} > \frac{\square}{\square}$$

THE MATHEMATICS

Student Observations

Where students will decide to place 9 or 0 on the first roll will vary for each game. For example, in *Game 1*, where the goal is the highest sum, most students will place 9 in the hundreds column. They usually place 0 in the ones column or discard it. For *Game 3*, where the goal is the highest product, students often put 9 in the tens column. Zero on the first roll is often discarded.

When rearranging the numbers to determine the best answer, students may find it helpful to use calculators to quickly and accurately try out different solutions.

Going Further

Have students create their own **Number Roll** game. They can use any math operation and decide the goal. Have them play some of these student-created games in small groups or as a class.

Place Your Values

2 players per team

2-3 teams

MATH SKILLS

place value, estimation

THE GAME

Getting Started

Players place numbers on a playing board, trying to create numbers that are as close as possible to target numbers. **Place Your Values** is an excellent game for developing place-value understanding and estimation skills.

Variations

Have students create their own **Place Your Values** games and play them with classmates.

Make one number on the decahedron a wild number. When it is rolled, players can choose any number they want to write in an empty box.

THE MATHEMATICS

Student Observations

When discussing their strategies, students often say that they try to fill in the lead numbers early in the game. For example, to make a number close to 500, students often wait for either 5 or 4 to write in the hundreds column.

Going Further

Ask students to determine the lowest total they can get with the numbers they rolled. To find out, have students rearrange the numbers on their playing boards. Students write their answers on a new playing board.

For more practice with place-value concepts, have students try these games: **Number Ladder**, **Number Roll**, **Position Power**, and **Place Your Decimals**.

Position Power

2 players per team
2-3 teams

MATH SKILLS

place value, estimation, addition

THE GAME

Getting Started

Position Power is a strategy game focusing on estimation and place value. Students write a number in either the ones or tens column in an attempt to create a five-number addition problem whose sum does not exceed 100.

Variations

For practice with subtraction, have students start at 100 and try to be the team closest to 0 after five rounds. Then follow the same rules.

Have students try the game with two octahedrons. Students will need a new playing board that includes the hundreds position. For example, a team who rolls 3 and 4 could write 304, 340, 430, 403, 43 or 34. The team closest to 1,000 after five rounds is the winner.

Challenge students to create their own rule variations. Some student-generated rules have included *write what you roll on your opponent's board* and *at the end of the game, change the place of one number*.

THE MATHEMATICS

Student Observations

Many students will say that if on their first turn they roll 5 or 6, they will write it in the tens column. This is because they feel they can always write the other rolls in the ones column and still have a number that is not over 100. If, however, they roll 7 or 8 on the first roll, they place it in the ones column.

Going Further

An interesting pattern emerges if the position of each number is reversed. Have students calculate their reversed **Position Power**

entries for each of their games. Give them plenty of time to discuss their results and ideas.

Students will discover the following pattern: When the final sum of the reversed-position entries is less than 100, this sum is the reverse of the **Position Power** sum. When the final sum of the reversed-position entries is greater than 100, the sum of the first and last digits is the tens place digit in the **Position Power** sum. The middle digit of the reversed number equals the ones digit in the **Position Power** sum.

Place Your Decimals

2 players per team

2-3 teams

MATH SKILLS

decimal place value, estimation

THE GAME

Getting Started

In **Place Your Decimals**, students write numbers on a playing board, trying to create inequalities involving decimals. In order to win, students must consider the place value of a number before writing it on the playing board.

Review the symbols for greater than (>), greater than and equal to (≥), less than (<), and less than or equal to (≤).

Variations

Make one number on the decahedron a wild number. When it is rolled, players can choose any number they want to write in an empty box.

Use a tetrahedron or a hexahedron to play the game. Fewer number options make filling in the board more challenging.

Have students make up their own **Place Your Decimals** games and play them with classmates.

THE MATHEMATICS

Student Observations

Students often fill in the leftmost boxes first in this game. When the leftmost numbers are correctly placed, the other places can be filled in with any numbers.

Going Further

Challenge students to rearrange numbers on their board to make all their math sentences true.

Have students play **Place Your Values**, a similar game which focuses on place value of whole numbers.

Wipe Out

2 players

MATH SKILLS

missing addends, addition

THE GAME

Getting Started

Wipe Out is a two-person game that focuses on creating different combinations of numbers whose sum equals the sum of the two numbers rolled on the dice. Players cross off one of these sums in a turn. The goal is to be the player whose crossed-off numbers have the highest sum.

Variations

Challenge students to try to get the lowest score possible. Discuss with them how this changes their strategy.

Have students use two different polyhedrons. Have them make new playing boards with numbers ranging from 1 to the highest sum possible with the polyhedrons selected.

THE MATHEMATICS

Student Observations

The lowest score possible is 2, which occurs if the student rolls double 1s two times in a row. The only number that can be crossed off with double 1s is 2. Since 2 is crossed off on the first turn, no other numbers can be crossed off on the second turn.

There are many different sequences of rolls that would enable students to cross off all fourteen numbers. The fewest number of turns needed is eight. This could happen if the sum of 14 is rolled seven times and the sum of 7 is rolled once.

Many students report that their strategy is to cross off the higher numbers first. They try to leave the lower numbers for making up com-

binations of numbers for sums already crossed off.

Going Further

Challenge students to determine all possible cross offs for a particular sum. For example, if the sum is 5, the following three combinations are possible: 5, 4 + 1, or 3 + 2. Have students find all the possible combinations for each sum. Ask: *Which sum has the most combinations? The fewest? What else did you notice?* Have students report on how they organized their work, which will vary among the groups. The chart below shows the number of combinations for each sum.

Sum	2	3	4	5	6	7	8	9	10	11	12	13	14
Combinations	1	2	2	3	4	5	6	8	10	12	15	18	22

Pig Out

2 players

MATH SKILLS

probability, addition

THE GAME

Getting Started

Pig Out is an easy-to-learn game. The goal is to be the first player to reach a score of 100 by finding the sum of the rolls of two octahedrons. Students can lose points, however, if they *pig out*. Rolling 1 on one or both octahedrons ends the student's turn and eliminates points.

Variations

For practice with subtraction, have students start with 100 points and try to be the first player to reach 0 following the same rules. Have students play **Pig Out** again, but this time each plays with a different polyhedron.

In each variation, have students consider how their strategy for winning changed.

THE MATHEMATICS

Student Observations

Students have a variety of strategies for winning **Pig Out**. Some students say that they will continue rolling the dice when their point total is small since there is less to lose if 1 is rolled. Other students say that they only take a few rolls per turn unless their opponent is nearing 100, in which case they will continue rolling as long as they can.

Students will develop a variety of different rules, such as adding 11 to the score, switching scores with the other player, and even interchanging the digits of the score whenever double 1s are rolled.

Going Further

Have students make an 8x8 addition chart showing all the possible sums that can happen with two octahedrons. Use this chart to show that the probability of rolling double 1s is 1/64 and that the probability of rolling 1 on either die is 15/64. Have students test these probabilities by rolling the two octahedrons until, for example, double 1s occur. Students should repeat the experiment several times and then find the average number of rolls it took to get double 1s. Have students compare their results to the expected probability.

Have students play this game with each of the following pairs of dice: decahedrons, icosahedrons, and tetrahedrons. After experimenting and comparing results, students realize that as the number of sides on the polyhedron increases, their chances of rolling 1 decreases.

Cross Out

2 players per team
2-3 teams

MATH SKILLS

mental math, addition, subtraction

THE GAME

Getting Started

Cross Out is a strategy game that incorporates addition and subtraction practice with problem solving. The goal is to start from a number between 40 and 60 and try to reach 0 exactly by deciding which numbers to cross out from the playing board.

Variations

Change the rules and play **Reverse Cross Out**. Students pick a target number, begin at 0, and add crossed-out numbers to try to reach their target number exactly.

Play **Cross Out** with different polyhedrons. Students can make up appropriate playing boards and set ranges for starting numbers.

THE MATHEMATICS

Student Observations

Students may say numbers closer to 40 are better starting numbers since there is less to subtract. Many students cross out the larger numbers early. In this way, they don't get stuck near the end of the game by not being able to cross out numbers needed to reach 0.

Going Further

Explore with students the fewest number of turns necessary to reach 0 if the starting number is 60. Have them list the possible rolls. It is possible to reach 0 in five rolls, and there are many solutions that will work for this.

Repeat this experiment using 40 as the starting number. In this case, it is possible to reach 0 in three rolls and, again, there are many different solutions that work.

Turning Products

2 players per team

2 teams

MATH SKILLS

factors, multiplication

THE GAME

Getting Started

Turning Products is a strategy game that involves logical thinking and knowledge of multiplication facts up to 12 x 12. The goal is to get four markers in a row horizontally, vertically, or diagonally. Students must carefully consider the factors of the numbers on the playing board to be sure they do not provide a winning move for their opponent's next turn.

Review the terms *factors* and *products*, as these are used in the rules. The numbers students multiply are the *factors*. The answer is the *product*. In the equation 2 x 3 = 6, 6 is the product, and 2 and 3 are the factors.

Provide two different kinds of beans or counters of two colors as markers.

Variations

Have students use any operation, not just multiplication. For example, if the dice were turned to 9 and 5, the team could use addition to place a marker on 14. Students must make a new playing board to include 0, 13, 17, 19, and 23 as these numbers can be made with addition or subtraction equations.

Change the goal so that the winner is the team that does not get four in a row. The strategy then is to force the other team to be the first to get four in a row.

Have students play with different polyhedrons and create their own playing boards.

THE MATHEMATICS

Student Observations

One strategy that students use in this game is to make a list of which factors they do not want to leave on the playing board for their opponent. Using this strategy, at some point in the game, one team will be unable to make a move that does not leave the other team with a winning factor since the available combina-tions are limited. The strategy for this game is similar to the strategy for **Gridlock**.

Going Further

Have students play the game several times with one change: The first team to get three marks in a row wins. Have them compare their results from both games. Discuss how their strategies changed.

All the Way to Ten

2 players per team
2-3 teams

MATH SKILLS

estimation, mental math, addition, subtraction, multiplication, division

THE GAME

Getting Started

Students use up to five numbers and any math symbols and operations they choose to create equations. The goal is to find solutions to these equations that equal the number of each round; that is, equal to 1 through 10.

This game is adaptable to a wide range of abilities. Since students will use math symbols and operations with which they are familiar, those less-experienced mostly tend to use the four operations. More-experienced students bring in more complex mathematical concepts such as square roots and exponents.

With the whole class, generate a list of all the different operations and math symbols that students know. Add symbols such as exponents, parentheses, decimal points, and square roots if students do not mention them.

In addition, let students know that they can put numbers together to make multidigit numbers when playing the game. For example, for a roll of 3, 5, and 9, they could use 35 and 9 in their equation.

Variations

For each roll, allow students to choose any number from 1 through 10 for the equation's answer. Once a number is used for an answer, it cannot be used as an answer again.

Have students use negative numbers, such as –1 to –10, for the round numbers.

Challenge students to use four 4s to create equations that equal the numbers 1 through 10. (Have them try it with four 3s or four 5s.)

THE MATHEMATICS

Student Observations

Have students create a class chart on which they record their solutions so that they can see the variety of possibilities.

Going Further

Discuss order of operations. Have students apply the rules to each equation on their playing boards. Have them discuss when the results changed.

Have students pick one of their rolls. Using these numbers, challenge students to find as many different ways to make equations that equal any or all of the numbers from 1-10.

Reach for Fifty

2 players per team

2-3 teams

MATH SKILLS

estimation, rounding, addition, subtraction, multiplication, division

THE GAME

Getting Started

Students create equations with the numbers and operations rolled on the polyhedrons. The goal is to come as close as possible to 50 after five rounds. As students play the game, they will need to estimate and make mental calculations as they determine the best equation to use.

Variations

Have students vary the goal for winning. Have them try to get the lowest score, the highest score, or the score closest to zero.

Allow students to change the rule for division. Have them express remainders as fractions or round quotients to the nearest hundredth.

THE MATHEMATICS

Student Observations

A strategy students often use is to create an equation whose solution is a high number (around 30) in the first round. Then, in subsequent rounds, they create equations with low-number solutions so that they slowly get closer to 50.

Going Further

For one roll of the polyhedrons, challenge students to determine the highest and lowest

possible answers in **Reach for Fifty**. Generally students report the highest answer as 516 from 86 x 6. If they use exponents, 6^8 x 6 yields 10,077,696; or $6^{(8 \times 6)}$ yields an even higher number, more than 20 undecillion (20 followed by 36 zeros). The lowest answer is −85 from 1 − 86. Students who are not familiar with negative numbers may report that the smallest answer is 0 from many different subtraction problems such as 1 − 1 or 4 − 4.

Make It Close

2 players per team
2-3 teams

MATH SKILLS

estimation, mental math, addition, subtraction, multiplication, division

THE GAME

Getting Started

Students create equations with numbers and operations rolled on the polyhedrons. The goal is to try to come as close as possible to their target number. **Make It Close** gives students practice with operations as they try to find the best equation.

Let students know they can put numbers together to make multidigit numbers when playing the game. For example, for a roll of 3, 5, and 9, they could use 35 and 9 in their equations.

Variations

Have students change the scoring so that if the equation equals the target number, the player scores 5 points. If the answer is within one number of the target number, the player scores 3 points; and if the answer is within two, the player scores 1 point.

THE MATHEMATICS

Student Observations

Students often report that they like rolling 1s, since they can make simple equations and alter them slightly by adding, subtracting, multiplying or dividing by 1.

The numbers 1, 2, 3, and 4 have a greater chance of occurring since they are the only numbers on all the polyhedrons. The numbers 0, 10, 11, and 12 occur only once on all five polyhedrons and may be rolled the least.

Using one of their rolls, students may be surprised to find that equations can generally be made to equal more than half of the numbers from 1 to 20.

Going Further

Have students pick one of their rolls. Ask: *With the numbers and operations rolled, what is the mathematical equation that will result in the highest answer? the lowest answer?*

Ask students to play again, but this time with only the hexahedron, tetrahedron, or a combination of both. Have them compare their results to those of the original game.

Total Factor

2 players per team
2-3 teams

MATH SKILLS

factors, prime and composite numbers, addition, multiplication

THE GAME

Getting Started

In **Total Factor**, students roll four octahedrons, find their sum, find all the factors of the sum, and then find the sum of the factors. The sum of the factors represents their score for that round. The goal is to be the first team to reach a score of 400. **Total Factor** provides practice in finding factors of 4 to 32.

Variations

To practice multiplication, have students use two octahedrons, multiply the numbers, and find all of the factors of the product. Ask students to keep the target number of 400.

THE MATHEMATICS

Student Observations

Many students will create charts to record the point value for each sum. To find the highest-scoring roll, some students will eliminate sums from being high scoring because they know that the sum has only a few factors. Then they find the scores for the sums they think will have the highest scores. To find the lowest-scoring sum, students often begin with 4 and work their way up, quickly realizing that since each number includes itself as a factor, the lowest-scoring number has to be one of the first few numbers.

The lowest-scoring roll is a sum of 5, which scores 6 points. The highest-scoring roll is a sum of 30, which scores 72 points. It is not possible to score every number between 5 and 72.

All numbers with two factors are called *prime* numbers. Examples of prime numbers are 2, 3, 5, 7, and 11. Numbers with an odd number of factors are called *square* numbers. Square numbers have a factor that is multiplied by itself to equal the number; for example, 2 x 2 = 4 and 5 x 5 = 25. Challenge students to find all the square numbers from 1 to 200.

Ask students to determine the fewest number of turns a player will need to win. (Six is the minimum number if a sum of 30 is rolled each time.) Have students share the different ways to solve this.

Going Further

Have students research *perfect* and *amicable* numbers. These are names mathematicians have given to numbers based on the sum of the number's factors, excluding the number itself. The factors of perfect numbers, excluding the number itself, add up to the number. Six is a perfect number since its factors 1, 2, and 3 add up to 6. Two other perfect numbers are 28 and 496. Amicable numbers come in pairs; the sum of the factors of each number, excluding the number itself, equals the other number. The lowest pair of amicable numbers is 220 and 284.

Fraction Attraction

2 players per team

2-3 teams

MATH SKILLS

equivalent fractions, estimation, addition and subtraction of fractions

THE GAME

Getting Started

In **Fraction Attraction**, students roll three decahedrons, make fractions, and then find their sum. The goal is to have their sum closest to but not greater than 10 after five rounds.

The game provides practice in estimation. At the same time, it requires students to think about the relative size of fractions.

Variations

To practice subtraction of fractions, have teams begin at 10 and try to be the first to reach 0.

For a simpler version, have students use tetrahedrons or hexahedrons.

THE MATHEMATICS

Student Observations

A commonly reported strategy is the following. On their first or second turn, students try to create improper fractions that equal whole numbers, such as 4/1, 6/3, or 4/2. On subsequent turns, they create fractions close to 1.

Going Further

Have students order each set of fractions on their playing boards from smallest to largest. Or, ask them to sort their fractions: greater than 1/2, less than 1/2, equal to 1/2. Try the same activities with 1/3 and 1/4.

Ask students how many different fractions they can make from rolling three numbers and using only two of the numbers. Answers will vary. If all three numbers are different, such as 2, 3, and 5, then six fractions are possible: 2/3, 2/5, 3/2, 3/5, 5/2, and 5/3. If two numbers are the same, such as 2, 4, and 4, then three fractions are possible: 2/4, 4/2, and 4/4. If you add the challenge that students can use all three numbers, then 18 fractions are possible if the three numbers are different. For example, using 2, 3, and 5, these fractions are possible: 2/3, 2/5, 3/2, 3/5, 5/2, 5/3, 23/5, 32/5, 25/3, 52/3, 35/2, 53/2, 2/35, 2/53, 3/25, 3/52, 5/23, and 5/32.

Close Call

2 players per team
2-3 teams

MATH SKILLS

equivalent fractions, estimation, addition and subtraction of fractions

THE GAME

Getting Started

Close Call is designed to provide students with practice in addition and subtraction of fractions. The goal is to create equations that come close to a target fraction. Therefore, students must consider the relative values of the fractions and do some estimating.

Variations

Challenge students to use a blank hexahedron labeled with multiplication and/or division signs to practice multiplication and/or division of fractions.

Have students make their own fraction cubes by labeling a blank cube with a different set of fractions.

THE MATHEMATICS

Student Observations

One strategy students often use is to first make one fraction that is close to or equivalent to the target fraction. Then, using the remaining numbers, they try to make a very small fraction that is either added to or subtracted from the first fraction.

Going Further

Using two or more of the numbers rolled on the dice, have students make a fraction that is as close as possible to the target fraction. For example, if the target fraction is 1/4 and the numbers rolled are 4, 6, 7, 8, and 13, possible fractions close to 1/4 are 87/413, 4/13, and 13/48. Students can then calculate the difference between their fraction and the target fraction.

Gridlock

2 players

MATH SKILLS

coordinate graphing

THE GAME

Getting Started

Students strategically place markers on a coordinate grid. The goal is to get four markers in a row horizontally, vertically, or diagonally. The decahedrons generate the coordinates. On each turn, the player creates a new point by turning only one of the decahedrons. The game requires players to think ahead in order to either block their opponents from making rows of four or to set up the decahedrons so that they can make a row of four.

Review the meaning of *coordinate* and the meaning of a row *horizontally, vertically, or diagonally*. Review how to locate points on a coordinate grid.

Variations

Incorporate coordinate graphing in four quadrants by using a different playing board and four dice:

1 blue decahedron, numbered 0-9
1 red decahedron, numbered 0-9
1 blue hexahedron, labeled +, +, +, −, −, −
1 red hexahedron, labeled +, +, +, −, −, −

In this game, each coordinate has a number and a positive or negative sign. For example, if the red dice show 3 and +, and the blue dice show 4 and −, then the point is (+3, −4). Follow the same rules as before, but this time, turn two dice — one number and one sign.

THE MATHEMATICS

Student Observations

Often students will make a list of the coordinates that will enable their opponents to win. Then they try not to leave numbers that will enable their opponent to make these coordinates. If two students play this way, at some point one will be forced to leave the other with a winning number since the available combinations are limited. The strategy for this game is similar to the strategy for **Turning Products**. A winning strategy when playing three in a row is to be the person who goes first.

Going Further

Change the rule so that the player who does *not* get four in a row wins. Have students compare winning strategies to those for four in a row.

Instead of getting four markers in a row, have students be the first to make a square of any size or orientation.